POND LIFE

Gerald Cox

Illustrated by
Howard Hanson

Michael Kesend Publishing, Ltd.
New York

Cover and book design by Jackie Schuman

Copyright 1988 © Gerald Cox

Dedicated to Lynn

CONTENTS

DAWN

Dawn

The first grays of dawn wash away the cool points of starlight and reveal the mirrored pond surface. At first light, the scene seems devoid of animal life. Creatures are just beginning to make cautious greetings to the dawn. Quiet short calls and muffled movements signal the waking of life around the pond.

The early morning glassy surface of the pond is for some creatures a floor and for others a ceiling.

Surface Tension

The surface of the pond is like a waterbed but we cannot really see or feel the skin-like surface where molecules of water attract each other and produce an invisible elastic skin. Creatures with fine hairs on their feet can walk on the skin and only dent it. Other animals hang from the underside of the skin like a spider on a ceiling. For some creatures, being caught under the skin can mean death.

From season to season, from dawn to dark, life in and around a pond is busy and ever-changing. Creatures are attracted to these shallow bodies of water by the abundant food produced in and around the area.

Some ponds are temporary and

creatures must hurry to complete their life cycles before the drying water brings their development to a halt. Other ponds are permanent and the community of life they support is more complex.

A pond with all its plants is like a solar collector. It generates a "boom or bust" pattern of life which relates to the plants' oxygen and temperature changes in the small body of water.

These shallow pools occur in many environments such as deserts, steamy southern bottomland, frigid mountain meadows or the midwest with its changeable seasons. Each pond attracts some of the same creatures, yet is a generator and magnet of life unique to the environment of the area.

To understand pond life, it is necessary to look at it as a whole. As with any community, human or natural, there are members that produce the food. In a pond, food production begins with the green plants. They are the basic link in the food web of the community upon which other plant-eating animals feed. These animals are then eaten by other organisms. So the food web builds.

Common names of living things can cause a great deal of confusion. The use of scientific names avoids much of this problem. A snake may be called a garter snake in one area, a ribbon snake in another area and a water or garden snake in still another area. Students of biology know exactly what animal is being discussed when the genus *Thamnophis* and the species *sirtalis* are used. It is an Eastern garter snake and all biologists the world over know this because all agree to use the same scientific name for that animal. No other animal has that scientific name.

GREEN PLANTS PRODUCERS

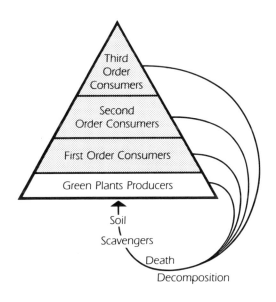

Green Plants

Essentially all the oxygen living things use comes from green plants. As sunlight strikes the green chlorophyll within the leaf or stem, it changes the carbon dioxide and water inside the leaf and converts them into oxygen and carbohydrates.

Green plants also provide the basic building block of the food system of a pond. The plants link one group of

animals directly or indirectly to another.

Within the water of a pond, green plants may range from microscopic to fairly large in size. Generally, the smaller the plant, the smaller the animals which eat it. There are exceptions to this general rule which you will discover as your study of a pond continues.

Plant and related animal life outside the pond but near the water are also linked directly or indirectly to the food system of the pond.

Algae

This group of plants is very basic to the food system of a body of water. Often they are thought of in a negative way because of the scum-like appearance they sometimes give to water. Pollution or nutrient enrichment of water may cause algae to increase rapidly and this creates water quality problems.

Some algae form thick green thread-like mats while others are single cells which float around.

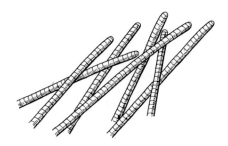

Blue-Green Algae

These are the simplest of all green plants. The chlorophyll is not enclosed within chloroplasts (special structures which usually enclose the chlorophyll). Some scientists say algae most nearly represent the earliest plant forms.

Look for these as single cells or in chains which are usually surrounded by a sticky jelly.

Blue-green algae can be found nearly everywhere: ponds, streams, lakes, sewers and ditches. If there are large numbers in drinking water reservoirs, the water may be considered tainted.

Green Algae

This type of algae is bright green and has the chlorophyll enclosed within chloroplasts. Spirogyra is a thread-like type which will form thick mats. Chlorella occurs as single cells.

Volvox is still another green algae. These microscopic forms look like hollow balls which seem to roll through the water.

Golden Algae

These algae, also known as diatoms, come in fantastic and beautiful shapes. They do not have cellulose cell walls like other algae but rather a shell made of glass-like silica.

Diatoms also produce food as do other algae. Much of their extra food is stored as oil.

If you are looking for ideas for jewelry designs, take a microscope and head for a pond to observe and sketch diatoms.

Euglenoids

These algae seem to be part plant and part animal. They contain chlorophyll and chloroplasts as do green plants. They move by means of a thread-like "whip" and have a primitive eye which cannot see and a throat-like structure (gullet) which takes in food.

Millions of these in a pond will make the surface water appear green.

Floating Plants

At first glance, the plants most noticed within the pond are those floating on the surface. Bubbles in the plant cells float the leaves yet the stems reach to the roots in the bottom of the pond.

Floating (and submerged) water plants help make water clear and add oxygen to the water. Aquatic animals not only eat the plants but also find shelter around the stems and leaves.

Water Lilies
Nymphaea tuberosa

There are many kinds of water lilies growing throughout the country. Beautiful and interesting flowers make this group really stand out. Leaf size ranges from two to twenty-four inches across.

Many pond creatures lay their eggs under the leaves and on the stem. The large leaves are used as resting areas for these creatures. Moose, beaver and muskrat are just some of the larger mammals that feed upon the water lilies.

Duckweed

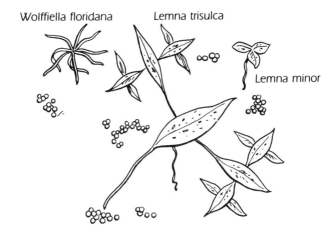

Wolffiella floridana   Lemna trisulca

Lemna minor

Some quiet ponds are nearly covered
with a green blanket of duckweed.
Many people are repelled by this scum
on the water in mid to late summer.
Upon closer examination, the blanket is
found to be made of individual plants.
They are some of the smallest flowering
plants in the world. These green
producers in the pond food web
provide for several varieties of water
birds and are a habitat for minute
organisms.

Water Hyacinth

Eichhornia crassipes

Water hyacinth is very attractive with its violet-white flower. It is, however, not native but an import from South America. Since few North American animals eat it, waterways and ponds are becoming choked with the plant, much to the dismay of boaters.

Small aquatic mammals do find shelter in and around the plant.

Water Lettuce

Pistia stratiotes

Water lettuce is found in ponds of the warm southeastern parts of the United States.

It covers the pond making it to look like a flat lawn. Alligators may be seen quietly lounging amid the floating plants.

The plants reproduce by sending out runner-like buds and may become stranded upon the shore or shallow bottom and take root.

Arrowhead

There are many species in this group of floating plants. Not all have arrowhead-shaped leaves, nor do all of them float.

Look for three-petaled whitish flowers sticking out

Sagittaria latifolia

Sagittaria graminae

of the water. Other names for this group are water plantains, wapato and duck potatoes. The tubers under the mud are excellent food for ducks and some mammals. Indians and pioneers also ate the tubers.

Coontail
Ceratophyllum dermersum

Coontail, also called hornwort, is one of the most common submerged plants. It does not have roots and seems to grow well when floating or anchored. In some ponds, it may become so thick that other aquatic plants are crowded out.

Ducks are the major consumers of the seeds of this plant.

Parts of the plants will fall to the bottom of a pond, become the new plants of spring and go on to produce oxygen.

Coontail can be seen in aquariums.

Watermilfoils

Myriophyllum heterophyllum

There are several varieties of watermilfoils scattered throughout the country. They are commonly used as aquarium plants.

These submerged, delicate plants are food for ducks but not to a great extent.

In the spring, new plants develop from the buds which wintered in the mud of the bottom.

Wild Celery
Vallisneria americana

This is a big favorite of several species of ducks. Diving ducks will go after the tender shoots which emerge from the roots in the bottom. Other ducks will often pick up the leaf bits floating on the surface.

In some areas, this is called eelgrass or tapegrass. The narrow leaves can become quite long, float on the surface and undulate like eels or floating tape.

Waterweed
Anacharis canadensis

This common plant grows in ponds in the northern tier of the United States, where it may float as dense mats. There are three varieties, and few animals seem to eat any of them.

Waterweed is commonly used in aquariums and is called "elodea".

Pondweed

Potamogeton amplifolius

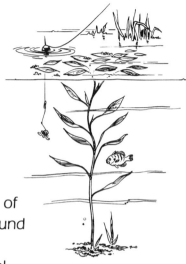

Several varieties of
pondweed abound
in this country.
Although several
are closely related, they often look
different. They are very widespread but
grow mostly in the northern and eastern
United States.

Seed, stems and tubers of pondweed
are all consumed by many ducks,
wading birds, muskrats and moose.

Like many other plants, new growth
each spring uses the starch stored in the
roots to provide energy for cell division.

Cattail

Typha latifolia

Two types of cattail exist: the
narrow leaf and broad leaf.
Broad leaf cattails are the
more abundant. Look for the
green shoots poking through
the water in mid-spring. Soon
the green male portion of the
flower will begin to develop
yellow pollen which then falls
downward to fertilize the female pistils
of the flower beneath it on
the same stem. This female
portion will become the
familiar brown cattail fluff.
Each particle of fluff is a seed
  Shelter, nesting material
and food are provided by
cattails. Muskrats are one of
the primary feeders and
users of these plants.

Horsetail

Equisetum fluviatile

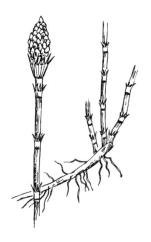

Horsetails are living fossils. Now they are three–feet tall with quarter–inch diameters, but millions of years ago they grew to the size of trees. These and other prehistoric plants exist today as coal beds and are being mined and burned to power our energy needs.

Years ago they saw much use as scouring pads for cleaning cooking pots. Silica in their cells acts as an abrasive.

They are commonly found in the shallows of ponds or growing on damp ground.

Bur Reed
Sparganium americanum

Bur reed is common in shallow ponds and is found more in the northern half of the country.

The burr-like seeds give the plant its name. It has several varieties and grows up to five feet tall.

Many animals use the bur reed plants to meet their food and shelter needs. Muskrats are probably the largest consumers but several ducks, shorebirds and even deer use it.

Carnivorous Plants

Pitcher plant
Sarracenia purpurea

Along the edge of the pond there exist some pretty aggressive plants. Insect-eating plants are truly something to behold. The pitcher plant resembles a vase about five inches tall. At the opening there are many downward-pointing hairs. When an insect lands on the slippery hairs, it falls into the pitcher and the hairs keep the insect from crawling out. In the bottom of the pitcher is a liquid which digests the insect, and the nutrients are absorbed into the plant cells.

Another insect-eating plant is the sun dew. The leaf is about the diameter of a dime and has sticky tipped hairs on the surface. If an insect crawls onto the leaf, hairs bend over and trap the prey. The insect is then digested on the surface of the leaf.

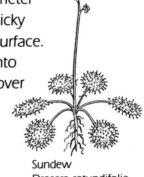

Sundew
Drosera rotundifolia

Venus Flytrap
Dionaea muscipula

Venus flytrap is another carnivorous plant. The modified leaves look like an open mouth with bristle-like teeth. When an insect trips the triggers inside, the trap slams and captures the prey, which is then digested.

FIRST ORDER
CONSUMERS

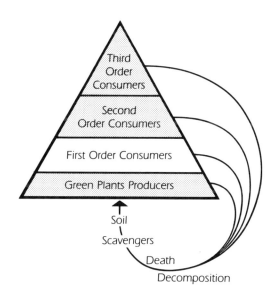

First Order Consumers

This consumer level is made up of creatures which are mostly plant eaters, although some may eat both plants and animals.

Competition for food at this level may occur at a microscopic level or the consumers may be quite large.

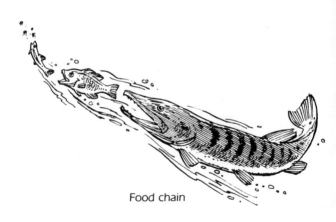

Food chain

Clams

If you find a furrow in the mud or sand on the bottom of a pond, it may lead you to a clam. Some freshwater clams are as large as five inches across the shell while others are the size of a person's little fingernail.

They move by extending a fleshy foot from the shell which pushes it along.

Food for the clam consists of tiny plants and animals which are sucked in, along with water, through a tube. When the water is discharged through another tube, the food is left behind inside the body of the clam.

Clams are a source of food for larger pond creatures.

Sponges
Spongilla fragilis

Sponges in a freshwater pond? They are usually thought of as living in salt water.

Look for a bright green mass surrounding a branch. The normal color of sponges is gray or brown but they are often green from the algae which grow on or within them.

Sponges are animals. They are a link between single-celled animal types and many-celled animal types.

When winter comes, the sponges disintegrate but some special cells fall to the bottom and become next year's sponges.

Protozoa

Paramecium caudatum

Euglena viridis

Amoeba proteus

A microscope is necessary to see the many different organisms of this group. Their numbers can be so great that a scum or cloudiness may form in the water and a fishy taste or odor may result.

Many move about under their own power by the use of hair-like structures and, while doing so, contact and eat smaller organisms and algae.

Protozoa, in turn, become food for small fish and animals that strain the water and ingest the protozoa.

Pond Snail

These pond creatures spend much time on the bottom of floating leaves or on the bottom of the surface film. There are two common shapes of snails. One is the pond snail which has a cone shaped, spiraled shell; the other is a flat-disk shape.

Snails move about on their slime-coated, muscular foot, eating algae and any other living or dead organic material with rasp-like tongues. Because they have a lung, they come to the surface to breathe.

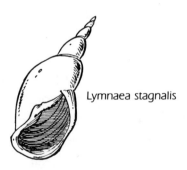

Lymnaea stagnalis

Planorbis trivolvis

Another group of common pond snails are those with flattened spiral shells. This group and the cone snails have tentacles and also move about on a slime-coated, muscular foot.

Try to find some snail eggs. Different types of snails lay differing egg masses. Most eggs are within a clear jelly. Some are in circular pads of jelly while others deposit jelly with eggs strung like beads inside.

Caddisfly
Rhyacophila fenestra

If you lie still and peer carefully into the pond's shallows you may see what appears to be a tiny stick moving across the bottom. Close examination of that stick will reveal a worm-like larva stage of the caddisfly inside with its head and legs extended. The larva glues tiny sticks or other materials together to make a well camouflaged temporary home. Soon the immature caddisfly will crawl out onto a reed, split open and fly away as a moth-like adult.

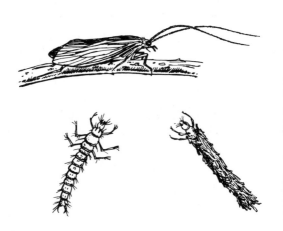

Mayfly
Hexagenia bilineata

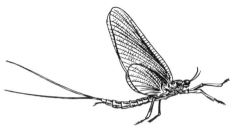

These delicate flying insects begin their lives as underwater nymphs. The immature stages are: nymph, larvae and pupae. Mayfly nymphs (there are several species) crawl on the bottom and cling to stems and rocks. They consume microscopic green plants and partly decayed plant material.

After transformation to adults from the nymph stage, they will fly away, complete a mating flight and mate. The male then dies and the female lives only long enough to lay eggs in the water.

As adults, they have no working mouth parts or digestive system. They live about twenty-four hours on stored energy from the nymph stage.

Wood Duck

Aix sponsa

This beautiful duck frequently is found
in or near a woodland pond. The male
(drake) is splendid with his iridescent
colors of green, blue, purple and brown.

The drake and hen quietly glide along
the shores eating seeds and water
plants. They also walk or fly into the
woods where they look for acorns,
berries, seeds and nesting holes.

Wood ducks nest in tree cavities.
Their eggs hatch in twenty-eight days;
then two-day-old ducklings plop ten to
twenty feet to the ground and the
parents escort them back to the pond.

Muskrat
Ondatra zibethica

A pleasant time can be spent observing muskrats. They are mostly nocturnal but can often be seen during the day carving gentle chevrons on the surface of the pond and then slipping quietly underwater.

Cattail, duck potatoes and sedges are their common plant food diet, but they will eat frogs, clams and fish.

A dome-shaped plant and mud house is a clue to their presence, although some dig burrows in mud banks.

Mink, raccoon, fox and hawks often make a meal out of a muskrat.

Moose
Alces alces

In the northern tier of the United States and Canada, this large member of the deer family can be found in a pond environment. Moose will browse in the vegetation around the pond but they also swim well and eat both floating and submerged aquatic plants.

If you find them feeding with heads submerged, that is a good time to stalk them for a closer look. Stay absolutely still when they lift their heads out of water. Caution is advised. Don't get too close because a frightened moose could do considerable damage to a watercraft.

SECOND ORDER CONSUMERS

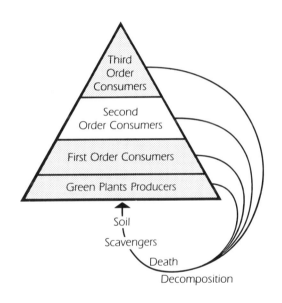

Second Order Consumers

Consumers at this level eat the animals which eat plants. For instance, a frog eats a plant-eating insect. However, this cannot be considered a hard–and–fast rule. Many animals will consume food one or two levels below their own on the pyramid.

There are exceptions to many rules in nature but generally the higher the level on the pyramid, the larger the consumer. Moose are a good example of an exception because they are plant eaters but also one of the largest animals in North America.

Mosquitos

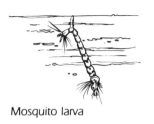

Mosquito larva

The adult mosquito is most noticeable around the pond environment. The female will extract a quantity of blood from any warm– blooded creature. Blood is necessary for the female to produce eggs. Males survive by sucking plant juices.

Mosquito eggs hatch after one to five days and then become larvae, which hang under the surface skin of the water eating microscopic plants and animals. After one to two weeks the larvae become pupae. At the pupa stage the adult form develops and then emerges to fly away.

Adult mosquito

Painted Turtle
Chrysemys picta

Painted turtles are the most common pond turtles. The young are a nice olive-brown on the back with the edge of the shell showing green or yellow. The bottom of the shell is usually a bright yellow and the head is black with yellow stripes.

Usually these turtles are sunning on a log and at times they are stacked one upon another.

Tadpoles, salamanders, snails, worms and frequently water plants are common food for this turtle. It will eat only when underwater.

Garter Snake

Thamnophis sirtalis

There are many species of garter snakes and they can be found in nearly all states from low elevations to mountain ponds.

Look for a two- to five-foot snake with two to three full length light stripes and dark scales between.

Garter snakes seem at home in or out of water and eat many animals including small mammals, birds, amphibians, and worms.

Careful handling will prevent a bite and a musky discharge on your hands. They become tame quickly.

Common Water Snake

Natrix sipedon

Probably this is the snake you will see most often in the central and eastern parts of the United States. It is about three- to four-feet long with a heavy body. The color pattern is light brown to reddish blotches on a light background.

Often this snake will wait on a tree branch over water basking in the sun and will dive into the water to capture a small fish, frog or crayfish.

Water Moccasin

Agkistrodon piscivorus

Cottonmouth is another name for this inhabitant of the pond environment. It gets its name because, when disturbed, it will open its mouth and display a very white mouth lining.

At first glance it looks a bit like the common water snake, but it is usually longer and heavier.

Move carefully in the southern states where they are common, for these poisonous snakes do not give warning before they strike.

They feed in much the same manner as the common water snake except they capture larger prey.

Amphibians

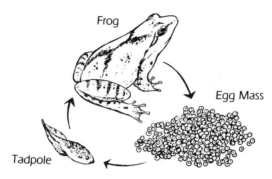

Typical frog life cycle

Most frogs and other amphibians must stay near water. Their skin allows water to quickly pass through it and the animal must stay wet or it will quickly dry out and die.

Reproduction in amphibians also requires water because their eggs lack shells; the eggs will quickly dehydrate unless submerged in water.

Frogs, toads and salamanders have similiar life cycles. The adults lay eggs in jelly-like strings or masses which hatch into tadpoles (larvae in salamanders) and then become adults.

Frogs

Of course, there are many different species of frogs. Some complete their life cycle in one season, others require two.

Bull frog
Rana catesbeiana

Largest of all is the bullfrog, which may reach eight inches in length. Look for this in the southern tier of states. The adult bullfrog is carnivorous, eating a variety of small animals.

The green frog is also called the pond frog. Its head and back are green and it is about three–inches long.

Green frog
Rana clamitans

Probably the most common and best known frog is the leopard frog. Look for a green body with yellow blotches.

Leopard frog
Rana pipiens

Early spring is announced by the spring peeper, a one-and-a-half-inch-long resident of woodland ponds. The loud shrill call would cause one to expect a larger frog. The body is brown with a dark X-mark on the back.

Toads

While toads are not pond dwellers as such, they are often found nearby. Because they are amphibians, toads must still return to water to reproduce. The skin of the toad is far more resistant to drying than the skin of the frog so the toad will venture farther from water.

The American toad is a very common sight in the eastern United States. Its brownish skin with lighter spots is characteristic. A warty appearance is also a sign for identification. The unique call is a loud trill.

American toad
Bufo americanus

Mid-continent
is the range of the
Great Plains toad.
Look for this insect
eater around
temporary ponds
and the sloughs
of the prairies. It is

Great Plains toad
Bufo cognatus

similiar to the American toad but its skin
has darker blotches on a lighter
background.

The western toad is found farther
west. Look for a brownish to greenish
toad. There are several members in the
western toad group, which accounts for
the color variation.
These toads also
use the temporary
ponds and
ditches for their
mating, and their
calls are heard at
that time.

Western toad
Bufo boreas

Salamanders

Expect to find a variety when observing this group of animals. Note that lizards have scales, claws and ear openings and salamanders have none of the above.

Lesser siren
Siren intermedia

Sirens are one representative group. They live their whole lives in water and eat crayfish and other aquatic plants and animals.

A common salamander is the tiger salamander. Look for a brown body with a yellow bar pattern. These will breed in small ponds and will become land dwellers when they lose their gills and become adults.

Tiger salamander
Ambystoma tigrinum

Backswimmer

Notonecta undulata

Aquatic insects called
backswimmers also
use the surface
film, clinging to the
underside. When
they are disturbed,

you will see them "row" toward the
bottom with jerking movements. They
swim on their backs using rowing legs.

Food for these consumers is often
small minnows, tadpoles and other tiny
prey.

When diving, they will carry a bubble
trapped among the body hairs. After a
couple of minutes, they will rise to the
surface to catch more air.

Caution: the backswimmer has a
sharp, needle-like mouth for piercing its
prey and may use it to give a painful
sting if you pick up the insect.

Whirlygig Beetle
Dineutes americanus

These are familiar dwellers of the surface environment. Graceful circles created by groups of these social insects are the sign of whirlygig beetles.

When approached closely, they will dive; when caught, they will give off a smelly fluid.

They float partly above and partly below the surface, and their divided eyes allow them to see at both levels. This lets them watch for insects to eat and at the same time watch for enemies below.

View of divided eyes

Water Strider

Gerris marginatus

One may easily spot the quick, jerking movements of water striders as they skate on the surface tension of the pond.

These half-inch to inch-long insects hunt other creatures which share the surface film of the water.

Water striders are capable of launching themselves into the air to capture prey.

Giant Water Bug

Lethocerus americanus

These giants of
the insect world
are huge
predators by
pond insect
standards; they
are one to
two inches,
depending on the species.

Stalking around the pond, they look
for tadpoles, minnows or any other
animal food they might grab with their
front legs.

Giant water bugs have a powerful
bite, even to humans.

Bat

Myotis lucifugus

Bats are truly nocturnal and are attracted to the location of their food, insects. What better place for bats to hunt than near a pond.

Look for bats swooping toward the pond surface at dusk. They use radar-like sound reflections to locate and catch their prey during their flight.

During the day, bats will roost under tree bark, in hollow trees or in rock cracks.

Bats are very important in the ecological balance because in one evening they eat many insects.

THIRD ORDER CONSUMERS

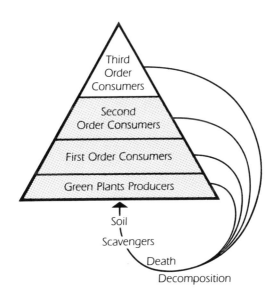

Third Order Consumers

Generally, consumers at this level are eaten by few other animals. They often die of old age unless death comes from the result of human activity. Animals in the third order tend to be large, live in inaccessible areas and defend themselves well. As with all consumers in an energy pyramid, there is overlap between the levels and roles are not always clearly defined.

In the end, when death comes, their bodies are decomposed and recycled by the scavengers.

Raccoon
Procyon lotor

Raccoons are most common near a pond with a woodland nearby. They prefer hollow trees for shelter.

Nearly everyone is familiar with their masked faces and tails with black rings.

These animals will eat nearly anything and swim well. Crayfish and frogs are common foods and they are quick to take the eggs or young of birds which might be nesting in the cattails.

Kingfisher

Belted Kingfisher
Megaceryle alcyon alcyon

Once heard, the rattling call of a belted kingfisher is not forgotten. This efficient hunter can be seen perching on a branch over the water watching for minnows, frogs, tadpoles and other creatures. Once prey is spotted, the bird dives forcefully into the water to capture a meal.

If there are dirt cliffs nearby, look for kingfisher tunnel nests there.

Great Blue Heron

Along the edge of a pond, this graceful bird can be a study in slow motion. It spends most of its time standing motionless waiting to sight a frog, crayfish or fish. A quick stab with its long pointed bill and it has a meal, which is usually swallowed whole.

This visitor to the pond will move with slow, graceful wing beats to hunt in another quiet backwater environment. You will not want to miss this four-foot-tall, gray bird.

Hawks

Marsh Hawk or Harrier
Circus cyaneus hudsonius

As the marsh hawk flies away, look for a white rump patch, rounded wing tips and a long tail. This bird is also known as a harrier.

Redtail hawks have a short body and tail and broad wings. This bird tends to soar and circle quite high while watching for prey, whereas the marsh hawk tends to stay low over the wet areas while hunting.

Snapping Turtles
Chelydra serpentina

Nearly everyone has heard about this legendary creature. Rarely will it be seen basking in the sun; usually it is moving about or resting on the bottom. It is the largest turtle—about fifteen inches in diameter in the north and larger in the south. Thirty-five pounds is large in the north; in the south it may be much larger.

Look for the gray rough shell and a tail with sharp ridges. Be cautious of the vicious mouth; it is quick to bite and surprisingly fast.

It will eat nearly any food it can catch.

Alligator

American alligator
Alligator mississipiensis

Expect to find this holdover from the dinosaur age in the lower Mississippi valley and the southeast coastal areas.

Look for these reptiles quietly resting on the shore or with eyes, snout or head sticking up above the water. They may be in ponds made by a "gator" itself. These pockets of water are very important to all wildlife during the dry season. Along the bank, look for the nest made of piled-up vegetation. Be cautious around the nest; the female will protect it with aggressive behavior.

SCAVENGERS

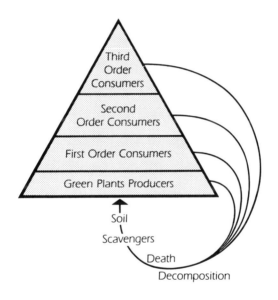

Scavengers

It is essential to all natural systems to keep the body chemicals of plants and animals available so they may be re-used by other organisms living in the environment.

Once a plant or animal dies, the chemicals, such as iron, potassium and calcium, are released by the rot and decay process. Those chemicals then leach into the soil and are taken into the plant roots to make new cells which are eaten by animals. In that way, chemicals are recycled within a natural system.

Many organisms assist in this movement: bacteria digest cells; fungus helps to break up vegetation to make humus and release chemicals; worms and insects eat the organic material making cells and these are eaten by other animals higher up in the food pyramid. This cycle has been a continuous process through millions of years. All life is interrelated, directly or indirectly.

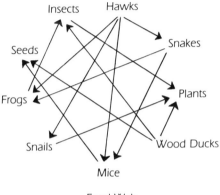

Food Web

Elimination of part of a food web through pollution or habitat destruction can drastically upset an entire ecological system.

Planaria
Dugesia tigrina

Although not microscopic in size, the smaller common types of planaria (flatworms) can best be observed with a stereoscopic microscope. Many flatworms are one-fourth of an inch long; some may reach one-and-a-half inches.

These creatures spend much of their time on the undersides of leaves and stones. Most of their food is animal, which they eat by extending their tube-like mouths into the food source.

Collect these by placing a small piece of raw liver at the edge of the pond. The worms will collect on the liver and can then be transferred to a jar.

Leech
Macrobdella decora

Many people shudder at the sight of a leech. Leeches are really very interesting animals. There are several varieties and most live in the plant–remains on the bottom of ponds. Not all are blood suckers of humans, but they will attach themselves to nearly any animal. Some are about one-and-a-half-inches long and others are up to ten inches in length.

Look for their graceful undulating swimming motion; some move by "looping" from one object to another (tail to head, tail to head).

Waterboatman
Arctowrixa alternata

At first glance, the waterboatman looks like the backswimmer. A closer look reveals it does not swim on its back but rows along with a jerking movement like the backswimmer. An air bubble is carried along as it dives to feed by sifting through the bottom ooze with its front legs.

This scavenger serves in an important niche by eating the plant and animal remains on the bottom.

Crayfish
Procambarus blandingi

Crayfish (called crawdads in some areas) are fascinating creatures. There are about two hundred species and not all live in ponds.

The best time to find them is at night so a flashlight is most useful.

Crayfish eat nearly anything they can glean off the bottom of a pond.

Be quick with a hand or net because they swim backward in very fast jerks. They are fun to watch in an aquarium.

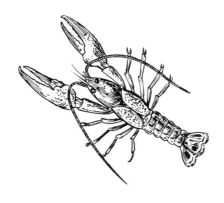

EQUIPMENT

Expensive gear is not necessary to begin a study of pond life. Most of it can be gathered around your home or can be easily made.

The best piece of equipment is a pair of eyes used to their fullest. Eyes and a powerful sense of curiosity make an excellent combination.

Nets are very useful tools. Insect nets may be made from netting from a fabric store. Sew it into a tube shape with one end tied shut. Keep one end open with a bent coat hanger and tie it to a stick for a handle.

A net to catch the very tiny plankton can be made by stitching a nylon stocking to a coat hanger on a stick.

Aquatic nets can be made as well. A net for obtaining bottom samples can be made by fastening a kitchen strainer to a long handle.

To make a throwing plankton net, use a knee high nylon stocking with a wire hoop and string on one end. Make a hole in the toe and tie a bottle at that end.

Wide–mouth jars of either glass or plastic in a variety of sizes are essential.

A white enamel cake pan is very useful for observing samples. The light background makes it easier to see the animals.

To gather plants which are out of reach, it is handy to have a throwing hook. This could be made with a variety of materials. Try a weighted board with nails pounded into it or bend some heavy wire into a hook and attach a string.

Eye droppers, tweezers and a magnifying glass are all especially useful.

A microscope would be very useful and many low-cost ones are now being sold.

As your interest and knowledge increase, better equipment and reference books will be a must.

Aquariums will allow long–term temporary or permanent observation of individual organisms. Trying to establish a balanced pond environment in an aquarium is an interesting project.

If a commercially made tank is available, that is excellent. Small aquariums can be made by cutting the top off a two-liter, clear plastic pop bottle, which is a good way to recycle them.

RECORD KEEPING

If you are serious about pond study or any nature study, a notebook becomes important. It does not need to be formal but a page should be kept each day you are making observations.

Information should include: date, time of day, location, weather, descriptions, sketches and measurements of specific observations.

Date _____Time _____

Location _____

_____

Weather _____

Observation _____

_____

_____

_____

_____

INDEX